One hundred ways to Number theory -Volume 1

firas kaabi

July 2023

1 Introduction

In this book, multiple exercises on number theory are provided along with their solutions.

We aim to cover all areas of this discipline. The primary objective of this book is to provide an accessible introduction to number theory, covering essential concepts, techniques, and theorems. Through a combination of examples and exercises, readers will develop a solid foundation in number theory and gain the necessary skills to solve problems and explore advanced topics.

the difficulty level is indicated, ranging from easy to medium, difficult, and very difficult. The reader is given the freedom to choose the appropriate difficulty level from the 50 exercises presented in this first volume. The remaining 50 exercises will be presented in Volume 2.

With both volumes, the reader will have a total of 100 number theory exercises, hence the title of the book.

2 Prime numbers and divisibility

Exercise 1 (easy): Prove that there are infinitely many prime numbers.

Reminder: A prime number is a number which is divisible only by 1 and itself.

Solution. To prove that there are infinitely many prime numbers, we will use a proof by contradiction.

Assume, for the sake of contradiction, that there are finitely many prime numbers. Let's denote them as $p_1, p_2, \ldots, p_n$, where n is the total number of primes.

Now, consider the number $N = p_1 \cdot p_2 \cdot \ldots \cdot p_n + 1$. This number is constructed by multiplying all the known prime numbers and adding 1 to the result.

N is either prime or composite.

Case 1: If N is prime, then we have found a new prime number that is not in our initial list of primes, contradicting our assumption that there are only finitely many primes.

Case 2: If N is composite, it must be divisible by some prime number. However, none of the primes $p_1, p_2, \ldots, p_n$ can divide N, because dividing N by any of these primes leaves a remainder of 1. Therefore, N must be divisible by a prime number that is not in our initial list of primes, again contradicting our assumption.

In either case, we arrive at a contradiction, which means our assumption that there are finitely many prime numbers must be false. Therefore, there must be infinitely many prime numbers.

$\square$

Exercise 2 (easy) : Prove that $A = (n+1)^2$ and $B = n$ are coprime for all $n \geq 1$.

Solution. Let's assume there is a positive integer d that is a common divisor of A and B. This means that d divides both A and B hence divides $A - B^2 - 2B = 1$

Therefore, A and B are coprime for all $n \geq 1$. $\qquad\square$

Exercise 3 (easy):

find all possible integers n such that $\frac{5n+4}{n+1}$ is an integer.

Solution. $\frac{5n+4}{n+1}$ can indeed be simplified as $\frac{5n+5}{n+1} - \frac{1}{n+1}$, which is equal to $5 - \frac{1}{n+1}$.

For $\frac{5n+4}{n+1}$ to be an integer, $\frac{1}{n+1}$ must also be an integer. This means that $n + 1$ must divide 1.

Considering the cases:

Case 1: If $n + 1 = 1$, then $n = 0$. So, $n = 0$ is a possible solution.

Case 2: If $n + 1 = -1$, then $n = -2$. So, $n = -2$ is a possible solution.

Hence, the possible integers n such that $\frac{5n+4}{n+1}$ is an integer are $n = 0$ and $n = -2$. $\qquad\square$

Exercise 4 (Medium): (Mersenne primes)

$a, n \geq 2$, prove that if $a^n - 1$ is a prime number then $a = 2$ and n is a prime.

Solution. Since $a^n - 1 = (a - 1)(a^{n-1} + a^{n-2} + \ldots + a + 1)$, and $a^n - 1$ is a prime number, we can deduce that $(a - 1)$ divides a^{n-1}. Moreover, we know that $a - 1 < a^{n-1} + a^{n-2} + \ldots + a + 1$. Therefore, we can conclude that $a - 1 = 1$, which implies $a = 2$.

Now, let's consider $n = pq$. We have $2^n - 1 = 2^{pq} - 1 = (2^p)^q - 1 = (2^p - 1)((2^p)^{q-1} + \ldots + 2^p + 1)$. This means that $(2^p - 1)$ divides 2^{n-1}. Since $2^n - 1$ is a prime, we can deduce that $2^p - 1 = 2^n - 1$, which implies $p = n$, or $2^p - 1 = 1$, which implies $p = 1$.

As a result, we can conclude that the only divisors of n are n and 1, which implies that n is a prime.

$\square$

Exercise 5 (Medium):

1/ Let $d(n) = \sum_{k|n} k$, where n is a non-negative integer.
Prove that n is a perfect square if and only if $d(n)$ is odd.

2/ Prove that $\prod_{d|n} d = (\sqrt{n})^{d(n)}$.

Reminder: k divides n if and only if $n = pk$, where p is a non-negative integer.

Solution. 1/We know that $d(n) = \prod_{i=1}^{r}(\alpha_i + 1)$.

Recall that a product of integers is odd if and only if all its factors are odd. Hence,

$d(n)$ is odd $\iff$ $\forall i \in \{1, \ldots, r\}$, $\alpha_i + 1$ is odd $\iff$ $\forall i \in \{1, \ldots, r\}$, α_i is even $\iff$ n is a perfect square.

2/The idea is to group together the factors d and n/d. If the first factor is less than $\sqrt{n}$, the second factor is greater than $\sqrt{n}$. Specifically, if n is not a perfect square:

$$\prod_{d\mid n} d = \prod_{d\mid n, d<\sqrt{n}} d \times \prod_{d\mid n, d<\sqrt{n}} \frac{n}{d} = \prod_{d\mid n, d<\sqrt{n}} d \times n^{\frac{d(n)}{2}}.$$

If n is a perfect square, $n = m^2$, then:

$$\prod_{d\mid n} d = m \prod_{d\mid n, d<\sqrt{n}} d \times \prod_{d\mid n, d<\sqrt{n}} \frac{n}{d} = n^{\frac{d(n)}{2}}.$$

$\square$

Exercise 6 (Medium):

Let q be an integer. Find an interval of length q that does not contain any prime numbers.

Solution. Let $n \geq 1$ and $I = \{n!+2, n!+3, \ldots, n!+n\}$. Take an integer $m = n! + k$ in I. Then k is a proper divisor of m, since $1 < k < m$ and $k \mid k$, $k \mid n!$ since $k \leq n$, hence $k \mid m$. Thus, I is an interval of $(n - 1)$ integers that does not contain any prime numbers. If we choose $n = q + 1$, then we have answered the problem. $\square$

Exercise 7 (hard):

Let $n \geq 2$ be an integer and $S_n = \sum_{i=1}^{n} \frac{1}{i}$. Show that S_n is never an integer.

Solution. Consider an integer n and p the unique integer such that $2^p \leq n < 2^p + 1$. Then, every integer $k \leq n$, different from 2^p, can be written as $k = 2^u v$ where v is odd and $u \leq p - 1$. Thus,

$$n \sum_{k=1}^{n} \frac{1}{k} = \frac{1}{2^p} + \sum_{k \leq n, k \neq 2^p} \frac{1}{k} = \frac{1}{2^p} + \frac{N}{2^p} q$$

where q is odd. We can deduce that

$$S_n = \sum_{k=1}^{n} \frac{1}{k} = \frac{q}{2^p} + \frac{N}{2^p}$$

and the numerator is odd while the denominator is even.... Therefore, S_n is not an integer. $\qquad\square$

Exercise 8 (very hard Mind your eyes !!!):

Find all non-negative integer solutions to $2^a + 3^b + 5^c = n!$.

Solution. **Case 1:** $n < 0$

By inspection, the only solutions in this case are (a, b, c, n) in $\{(1, 1, 0, 3), (2, 0, 0, 3), (4, 1, 1, 4)\}$.

Case 2: $n \geq 6$

Claim: There are no solutions in this case.

Case 2.1: $a = 1$

Case 2.1.1: $c = 0$

We have $3 + 3^b \equiv 0 \pmod{9}$, which is impossible.

Case 2.1.2: $c > 0$

Considering the equation modulo 5, we have $2 + 3^b \equiv 0 \pmod 5$. This implies that $3^b \equiv 3 \pmod 5$, thus $b = 4b' + 1$. However, we have $2 + 3^{4b'+1} + 1 \equiv 0 \pmod 4$, which is impossible.

Case 2.2: $a = 2$

Case 2.2.1: $c = 0$

We have $5 + 3^b = n!$, but considering the equation modulo 3, this is not possible. **Case 2.2.2:** $c > 0$, $4 + 3^b + 5^c \equiv -3^b \mod 5$. Thus, $3^b \equiv 1 \mod 5$, then $b = 4b' + 3$. We have $4 + 3^{4b'+3} + 5^c \equiv 7 + 5^c \equiv 0 \mod 8$. Hence, $c = 2c'$. Therefore, $4 + 3^b + 5^{2c'} \equiv 1 + 2^{2c'} \equiv 2 \not\equiv 0 \mod 3$. Thus, there are no solutions.

Case 2.3: $a = 0$. There are no solutions considering the equation modulo 2.

Case 2.4: $a \geq 3$.

Case 2.4.1: $b = 0$. Then, $2^a + 1 + 5^c \equiv 5^c + 1 \equiv 0 \mod 8$, which is impossible.

Case 2.4.2: $b > 0$. Consider the equation modulo 3: $2^a + 2^c \equiv 0 \mod 3$. That means a and c have opposite parities.

Case 2.4.2.1: $c = 0$. Considering the equation modulo 8, we have $3^b + 1 \equiv 0 \mod 8$, which is impossible.

Case 2.4.2.2: $c > 0$. $2^a + 3^b + 5^c \equiv 2^a + 3^b \equiv 0 \mod 5$. Then, a and b have the same parity. $2^a + 3^b + 5^c \equiv 3^b + 5^c \equiv 0 \mod 8$. Thus, both b and c are odd, but they can't be since a and b have the same parity and a and c have opposite parities. Therefore, the only solutions are $(1, 1, 0, 3)$, $(2, 0, 0, 3)$, and $(4, 1, 1, 4)$. $\square$

Exercise 9 (easy):
prove that $n(n+1)(n+2)(n+3)$ is divisible by 24

Solution. $n, n+1, n+2, n+3$ are 4 consecutive numbers then one of them is divisible by 4 .

without any loss of generality let n be divisible by 4. $n, n+1, n+2$ are 3 consective numbers then one of them is divisible by 3 and $n+2, n+3$ are consecutive numbers then one of them is even.

Hence $n(n+1)(n+2)(n+3)$ is divisible by $3 \times 4 \times 2 = 24$

$\square$

Exercise 10 (Medium):
prove that the equation $x^3 - x^2 + x + 1 = 0$ has no rational solutions.

Solution. Assume there exists a rational solution $x = \frac{p}{q}$, where p and q are coprime integers with $q > 0$. Substituting x with $\frac{p}{q}$ in the equation and multiplying both sides by q^3, we get:

$$p^3 - p^2 q + p q^2 + q^3 = 0.$$

Since q divides $p^2 q - p q^2 - p^3$, we have q divides p^3. Therefore, $q = 1$ since p and q are coprime. By applying the same reasoning to p, we obtain $p = \pm 1$. However, 1 and -1 are not solutions to the equation, leading to a contradiction.

$\square$

Exercise 11 (Medium): 1/let a and b be 2 non-negative integers such that $\gcd(a, b) = 1$.

prove that $\gcd(a^2, b^2 - a^2) = 1$

2/let n be a non-negative integer , prove that $\sqrt{\frac{n}{n+2}}$ is irrational.

Solution. let p be a prime number such that p divides a^2 and p divides $b^2 - a^2$, then p divides $b^2 - a^2 + a^2$, which is equal to b^2. Therefore, p divides both b and a^2, implying that p divides both b and a. Consequently, p divides the $\gcd(a,b)$, which is equal to 1. This is impossible. Hence, $\gcd(a^2, b^2 - a^2) = 1$. 2/Let's prove it by contradiction assuming that $\sqrt{\frac{n}{n+2}} \in \mathbb{Q}$, we can write $\sqrt{\frac{n}{n+2}} = \frac{a}{b}$, where $\gcd(a,b) = 1$ and $b > 0$. Thus, $\frac{n}{n+2} = \left(\frac{a}{b}\right)^2$. This implies that $n(b^2 - a^2) = 2a^2$.

Using the result from the previous question, we have $\gcd(b^2 - a^2, a^2) = 1$. Since a^2 divides $n(b^2 - a^2)$, by Gauss' lemma, we conclude that a^2 divides n. Therefore, we can write $n = ka^2$ with $k \in \mathbb{N}^*$. Substituting this into the equation, we obtain $k(b^2 - a^2) = 2$.

However, since $b \geq a + 1$, we have $b^2 - a^2 \geq (a+1)^2 - a^2 = 2a + 1 \geq 3$. Hence, $k(b^2 - a^2) \geq 3$. You got the contradiction , didn't you ??? If you didn't it's time to get some new glasses !!! $\qquad\qquad\square$

Exercise 12 (easy):

Find out the remainder of the euclidean division of $\sum_{k=1}^{n} k$ by n.

Solution. Let S_n be the sum of the first n integers. It is given by $S_n = \frac{n(n+1)}{2}$. If n is odd, i.e., if $n + 1$ is even, then $\frac{n+1}{2}$ is an integer, and n divides S_n. The remainder is thus 0. If $n = 2k$ is even, then $S_n = \frac{2k(2k+1)}{2} = k(2k+1) = k \cdot n + \frac{n}{2}$. The remainder is thus $\frac{n}{2}$. $\qquad\qquad\square$

Exercise 13 (beginner): find out $\gcd(63, 90)$ using two methods.

Solution. Method 1 :

$63 = 9 \cdot 7 = \underline{3^2} \cdot 7$ and $90 = 10 \cdot 9 = 5 \cdot 2 \cdot \underline{3^2}$, therefore $\gcd(63, 90) = \underline{3^2} = 9$.

Method 2 : using euclidean algorithm

$$
\begin{aligned}
90 &= 1 \cdot 63 + 27 \\
63 &= 2 \cdot 27 + 9 \\
27 &= 3 \cdot 9 + 0
\end{aligned}
$$

Therefore, $\gcd(63, 90) = 9$ since it's the last remainder different from 0. $\qquad\square$

Exercise 14 (Medium):

1/ prove the identity $x^n - y^n = (x - y) \sum_{k=0}^{n} x^k y^{n-1-k}$

2/ using the first question , prove that 609 divides $5^{4n} - 2^{4n}$.

Solution. 1/

$$
\begin{aligned}
(x - y) \cdot \sum_{k=0}^{n-1}(x^k \cdot y^{n-1-k}) &= \sum_{k=0}^{n-1}(x^{k+1} \cdot y^{n-1-k}) - \sum_{k=0}^{n-1}(x^k \cdot y^{n-k}) \\
&= \sum_{k=1}^{n}(x^k \cdot y^{n-k}) - \sum_{k=0}^{n-1}(x^k \cdot y^{n-k}) \\
&= \sum_{k=0}^{n-1}(x^{k+1} \cdot y^{n-k-1}) - \sum_{k=0}^{n-1}(x^k \cdot y^{n-k}) \\
&= x^n \cdot y^{n-n} - x^0 \cdot y^{n-0} \\
&= x^n - y^n
\end{aligned}
$$

2/

$$5^{4n} - 2^{4n} = (625)^n - (16)^n$$

$$= (625 - 16) \sum_{k=0}^{n-1} (625^k \cdot 16^{n-1-k})$$

$$= 609 \sum_{k=0}^{n-1} (625^k \cdot 16^{n-1-k})$$

$$5^{4n} - 2^{4n} = (625)^n - (16)^n$$

$$= (625 - 16) \sum_{k=0}^{n-1} (625^k \cdot 16^{n-1-k})$$

$$= 609 \sum_{k=0}^{n-1} (625^k \cdot 16^{n-1-k})$$

Therefore, 609 divides $5^{4n} - 2^{4n}$.

$\square$

Exercise 15 (Medium):
prove that for any integer $n \geq 1$, $40^n \cdot n!$ divides $(5n)!$

Solution. by induction :

The property is clearly verified if $n = 1$ since $5! = 120 = 40 \times 3$. Assuming the property holds for rank n, let's prove it for rank $n+1$. We have $(5n+5)! = (5n)! \cdot (5n+1)(5n+2)(5n+3)(5n+4)(5n+5) = (5n)! \cdot 5(n+1)(5n+1)(5n+2)(5n+3)(5n+4)$ and $40^{n+1}(n+1)! = 40^n n! \cdot 40(n+1)$. Since, by the induction hypothesis, $40^n n!$ divides $(5n)!$, it suffices to prove that $40(n+1)$ divides $(5n+1)(5n+2)(5n+3)(5n+4)$. But $5n+1, 5n+2, 5n+3, 5n+4$ are four consecutive integers. Two of them are even, and one of those even integers is divisible by 4. Hence, 8 divides their product. $\qquad\square$

Exercise 16 (Medium): Given p a prime prove that $\sqrt{p}$ is not a rationnal.

Solution. We'll prove it by contradiction. Let's assume that $\sqrt{p}$ is a rationnal. then Given that $\sqrt{p} = \frac{a}{b}$ and $\gcd(a,b) = 1$ with $b > 0$, we can conclude that $p = \frac{a^2}{b^2}$. Thus, $a^2 = p \cdot b^2$. Since p/a^2, it follows that p divides a. Since p is a prime number, we have $a = p \cdot k$ for some integer k. Substituting this into the equation, we get $p^2 \cdot k^2 = p \cdot b^2$, which simplifies to $pk^2 = b^2$. This implies that p divides b^2, and therefore, p divides b. This leads to the conclusion that p divides $\gcd(a,b)$. However, this contradicts the fact that $\gcd(a,b) = 1$. Hence, we have a contradiction.
$\qquad\square$

Exercise 17 (Medium): (Wilson's Theorem)
Let p be an integer greater than or equal to 2. Show that:
$(p - 1)! \equiv -1 \pmod{p} \Rightarrow p$ is prime.

Solution. Let p be a natural number greater than or equal to 2.
Suppose that $(p - 1)! \equiv -1 \pmod{p}$. There exists an integer
a such that $(p - 1)! = -1 + ap$ (*). Let $k \in \{1, ..., p - 1\}$.
The equation (*) can be written as $k\left(-\prod_{j=k}^{p-1} j\right) + ap = 1$. By
Bézout's theorem, it follows that k and p are coprime. Thus, p
is coprime with all natural numbers in the set $\{1, ..., p-1\}$, and
therefore, p is a prime number.

$\square$

Exercise 18 (Medium): Prove that the summ of 5 consecutive perfect squares can not be a perfect square.

Solution. For $n \geq 2$, $(n-2)^2 + (n-1)^2 + n^2 + (n+1)^2 + (n+2)^2 =$
$5n^2 + 10 = 5(n^2 + 2)$. If $5(n^2 + 2)$ is a perfect square, then 5
divides $n^2 + 2$.

However, if $n = 5q+1$ or $n = 5q-1$, then $n^2+2 \equiv 3 \pmod{5}$.
If $n = 5q + 2$ or $n = 5q - 2$, then $n^2 + 2 \equiv 1 \pmod{5}$. If $n = 5q$,
then $n^2 + 2 \equiv 2 \pmod{5}$.

As a result, $n^2 + 2$ is not congruent to 0 modulo 5 for all
$n \geq 2$.

$\square$

Exercise 19 (Medium):
Let $p \geq 2$ be a prime number. Show that p divides $\binom{k}{p}$ for $k \in \{1, \ldots, p-1\}$.

Solution. Since $\binom{p}{k}$ is an integer, we know that $\frac{k! \cdot (p-k)!}{p!} = \frac{p!}{p \cdot (p-1)!}$. Now, since p is a prime number, we also know that $k!$ mod $p = 1$ and $(p-k)!$ mod $p = 1$. By Gauss's theorem, $k! \cdot (p-k)!$ divides $(p-1)!$, and therefore $(p-1)! \cdot \frac{k!}{(p-k)!}$ is an integer. In other words, p divides $\binom{p}{k}$.

$\square$

Exercise 20 (Medium): For $n \geq 1$, show that $(n+1)$ divides $\binom{2n}{n}$.

Solution. It is clear that $(n+1)(2n+1) = n\binom{2n}{n}$. Now, since $(n+1) \wedge n = 1$, we can conclude that $(n+1)$ divides $\binom{2n}{n}$.

$\square$

3   congruence

Exercise 1 (easy):

  1/Find $n \in \mathbb{N}$ such that $5^n \equiv -1 \pmod{13}$.

  2/Find $n \in \mathbb{N}$ such that 13 divides $5^{2n} + 5^n$.

Solution. 1/We calculate the first terms and find that: $5^0 \equiv 1 \pmod{13}$, $5^1 \equiv 5 \pmod{13}$, $5^2 \equiv -1 \pmod{13}$, $5^3 \equiv -5 \pmod{13}$, $5^4 \equiv 1 \pmod{13}$, $5^5 \equiv 5 \pmod{13}$, $5^6 \equiv -1 \pmod{13}$, ...

We clearly observe the repeating cycle: $1, 5, -1, -5, 1, 5, -1$. This suggests that we can prove by induction, for $p \in \mathbb{N}$, the following property $P(p)$: $P(p)$: "$5^{4p} \equiv 1 \pmod{13}$, $5^{4p+1} \equiv 5 \pmod{13}$, $5^{4p+2} \equiv -1 \pmod{13}$, $5^{4p+3} \equiv -5 \pmod{13}$".

The property $P(0)$ is true, as shown by the previous calculation. Let $p \in \mathbb{N}$ be such that $P(p)$ is true, and let's prove $P(p+1)$. We have: $5^{4(p+1)} = 5^{4p} \cdot 5^4 \equiv 1 \cdot 1 \equiv 1 \pmod{13}$.

The proofs for the other three cases are exactly the same. Therefore, $P(p+1)$ is true.

Thus, the natural numbers that satisfy $5^n \equiv -1 \pmod{13}$ are exactly the numbers of the form $4p + 2$, where $p \in \mathbb{N}$.

  2/Let's write $5^{2n} + 5^n = 5^n(5^n + 1)$. If $13 \mid (5^{2n} + 5^n)$, and since $5^{\phi(13)} = 1$ (where ϕ is the Euler's totient function), the theorem of Gauss guarantees that $13 \mid (5^n + 1)$, which means $5^n \equiv -1 \pmod{13}$. According to the previous question, this is equivalent to saying that $n = 4p + 2$, where $p \in \mathbb{N}$. Conversely, if $n = 4p + 2$ for some $p \in \mathbb{N}$, we know that $13 \mid (5^n + 1)$, and therefore $13 \mid (5^{2n} + 5^n) = 5^{2n} + 5^n$. The integers n that satisfy the equation are exactly those of the form $4p+2$ with $p \in \mathbb{N}$. $\square$

Exercise 2 (Medium):

prove that 13 divides $3126 + 5126$

Solution. We will start by studying the powers of 3 modulo 13. We observe that: $3^1 \equiv 3 \pmod{13}$, $3^2 \equiv 9 \pmod{13}$, $3^3 \equiv 1 \pmod{13}$.

Therefore, since $126 = 3 \times 42$, we have $3^{126} \equiv (3^3)^{42} \equiv 1 \pmod{13}$. Similarly, we have: $5^1 \equiv 5 \pmod{13}$, $5^2 \equiv 12 \pmod{13}$, $5^3 \equiv 8 \pmod{13}$, $5^4 \equiv 1 \pmod{13}$.

Using a similar reasoning and noting that $126 = 31 \times 4 + 2$, we find that $5^{126} \equiv 5^2 \equiv 12 \pmod{13}$.

From this, we deduce that $3^{126} + 5^{126} \equiv 0 \pmod{13}$, which means that 13 divides $3^{126} + 5^{126}$.

$\square$

Exercise 3 (Medium):

prove that for every natural number n, $32n + 1 + 24n + 2$ is divisible by 7.

Solution. Let's consider, for $n \in \mathbb{N}$, the property $P(n)$ defined as "$32n + 1 + 24n + 2$ is divisible by 7". We will prove by induction that for every $n \in \mathbb{N}$, $P(n)$ holds.

Initialization: For $n = 1$, we have $32 \cdot 1 + 1 + 24 \cdot 1 + 2 = 7$, which is divisible by 7.

Induction: Let $n \in \mathbb{N}$ be such that $P(n)$ is true. Notice that $32 \equiv 2 \pmod{7}$ and $24 \equiv 2 \pmod{7}$. Then, we can write (modulo 7):

$32n + 3 + 24n + 6 \equiv 32 \cdot 32n + 1 + 24 \cdot 24n + 2 \equiv 2 \cdot (32n + 1 + 24n + 2) \equiv 2 \cdot 0 \equiv 0 \pmod{7}$.

Thus, 7 divides $32n + 3 + 24n + 6$, and therefore $P(n + 1)$ is true.

In conclusion, by the principle of induction, $P(n)$ is true for every $n \in \mathbb{N}$.

$\square$

Exercise 4 (Medium):
prove that if $a^2 + b^2 \equiv 0 \pmod 7$, a and b integers , then $a \equiv 0 \pmod 7$ and $b \equiv 0 \pmod 7$.

Solution. Let's start by writing a table describing the squares modulo 7.

$x \mod 7$	0	1	2	3	4	5	6
$x^2 \mod 7$	0	1	4	2	2	4	1

We can then distinguish four cases: - If $a^2 \equiv 0 \pmod 7$, then $a^2 + b^2 \equiv 0 \pmod 7$ if and only if $b \equiv 0 \pmod 7$. This is the case where both a and b are divisible by 7. - If $a^2 \equiv 1 \pmod 7$, for $7 \mid a^2 + b^2$ to hold, it would require $b^2 \equiv 6 \pmod 7$, which is impossible. - If $a^2 \equiv 2 \pmod 7$, for $7 \mid a^2 + b^2$ to hold, it would require $b^2 \equiv 5 \pmod 7$, which is impossible. - If $a^2 \equiv 4 \pmod 7$, for $7 \mid a^2 + b^2$ to hold, it would require $b^2 \equiv 3 \pmod 7$, which is impossible.

Thus, the only possibility is that both a and b are divisible by 7.

$\square$

Exercise 5 (Medium):
prove that "n^2 divides $(n+1)^{n-1}$ for every $n \in \mathbb{N}^*$"

Solution. $(n+1)^n = \sum_{k=0}^{n} \binom{n}{k} \cdot n^k$, so $(n+1)^n - 1 = \sum_{k=1}^{n} \binom{n}{k} \cdot n^k = \binom{n}{1} \cdot n + \binom{n}{2} \cdot n^2 + \cdots + \binom{n}{n} \cdot n^n$.

Since $\binom{n}{1} \cdot n = n^2$ is congruent to 0 $(\mod n^2)$, and for $2 \le k \le n$, n^2 divides n^k, then $\binom{n}{k} \cdot n^k$ is congruent to 0 $(\mod n^2)$.

Hence, $(n+1)^n - 1$ is congruent to 0 $(\mod n^2)$. As a result, n^2 divides $(n+1)^n - 1$.

$\square$

Exercise 6 (Medium): Given p and odd prime number such that p divides $n^2 + 1$, n is a non-negative integer , prove that $p \equiv 1 \pmod 4$

Solution. If we assume that p divides $n^2 + 1$, then $n^2 \equiv -1 \pmod p$. If p is odd, we can express it as $p = 2q + 1$, where $q \in \mathbb{N}^*$.

If p divides $n^2 + 1$, then n is not divisible by p.

Using Fermat's Little Theorem, we know that $n^{p-1} \equiv 1 \pmod p$. Since $n^2 \equiv -1 \pmod p$, we have $(n^2)^q \equiv (-1)^q \pmod p$. Hence, $1 \equiv (-1)^q \pmod p$.

Since $p \geq 3$ and p divides $1 - (-1)^q$, and since $1 - (-1)^q$ can only be 0 or 2, we conclude that $(-1)^q = 1$. This implies that $q = 2q'$, and thus $p = 4q' + 1$.

As a result, $p \equiv 1 \pmod 4$.

$\square$

Exercise 7 (Medium):
Find out the units digit of 7^{98}

Solution. To find the units digit of 7^{98}, we observe the pattern in the units digits of powers of 7:

$7^1 = 7$ (units digit: 7), $7^2 = 49$ (units digit: 9), $7^3 = 343$ (units digit: 3), $7^4 = 2401$ (units digit: 1), $7^5 = 16807$ (units digit: 7), $7^6 = 117649$ (units digit: 9), ...

We notice that the units digit repeats in a cycle: 7, 9, 3, 1. Since the cycle has a length of 4, we can divide 98 by 4 to determine which position in the cycle the power 7^{98} corresponds to.

98 divided by 4 gives us a quotient of 24 with a remainder of 2. This means that the units digit of 7^{98} is the same as the units digit of 7^2, which is 9.

Therefore, the units digit of 7^{98} is 9.

$\square$

Exercise 8 (Hard):

prove that the equation $(E) : x^2 - 7y^2 = 3$ has no solutions in $\mathbb{Z} \times \mathbb{Z}$

Solution. To prove that the equation $(E) : x^2 - 7y^2 = 3$ has no solutions in $\mathbb{Z} \times \mathbb{Z}$, we will use proof by contradiction.

Assume that there exists a solution $(x, y) \in \mathbb{Z} \times \mathbb{Z}$ that satisfies the equation $(E) : x^2 - 7y^2 = 3$.

Let's consider the equation modulo 7. Taking both sides of the equation modulo 7, we have:

$x^2 \equiv 3 \pmod 7$

Now, let's examine all the possible remainders of x^2 when divided by 7:

$x^2 \equiv 0^2 \equiv 0 \pmod 7$, $x^2 \equiv 1^2 \equiv 1 \pmod 7$, $x^2 \equiv 2^2 \equiv 4 \pmod 7$, $x^2 \equiv 3^2 \equiv 2 \pmod 7$, $x^2 \equiv 4^2 \equiv 2 \pmod 7$, $x^2 \equiv 5^2 \equiv 4 \pmod 7$, $x^2 \equiv 6^2 \equiv 1 \pmod 7$.

From this analysis, we see that x^2 can only be congruent to 0, 1, 2, or 4 modulo 7. None of these congruences satisfy $x^2 \equiv 3 \pmod 7$. Therefore, there are no integer solutions $(x, y) \in \mathbb{Z} \times \mathbb{Z}$ that satisfy the equation $(E) : x^2 - 7y^2 = 3$.

Hence, the equation (E) has no solutions in $\mathbb{Z} \times \mathbb{Z}$.

$\square$

Exercise 9 (Medium):

1/prove that a number is divisible by 11 if and only if the difference between the sum of its digits in odd positions and the sum of its digits in even positions is divisible by 11.

2/Is 854555673 divisible by 11 ?

Solution. 1/ a is a non-negative integer. We can express a as
$a = a_n \cdot 10^n + \ldots + a_1 \cdot 10 + a_0 = a_0 - a_1 + a_2 - a_3 + \ldots + (-1)^n \cdot a_n$
mod 11. Then, 11 divides a if and only if $a_0 - a_1 + a_2 - a_3 + \ldots + (-1)^n \cdot a_n \equiv 0 \mod 11$. Additionally, we can say that 11 divides a if and only if $\sum_{k \equiv 0 \mod 2} a_k - \sum_{k \equiv 1 \mod 2} a_k \equiv 0 \mod 11$.

2/Let $n = 8554555673$ and consider the alternating sum of its digits. We have $s = 3 - 7 + 6 - 5 + 5 - 5 + 4 - 5 + 8 = 4$. Since s is not divisible by 11, it implies that n is not divisible by 11.

$\square$

Exercise 10 (Medium):

Solve the equations :

$$\begin{cases} 4x \equiv 5 \pmod 9 \\ 7x \equiv 8 \pmod 9 \\ 3x \equiv 6 \pmod 9 \end{cases}$$

Solution. To solve the congruence equation $4x \equiv 5 \pmod 9$, we can find the multiplicative inverse of 4 modulo 9 and multiply both sides of the congruence by this inverse.

In this case, the multiplicative inverse of 4 modulo 9 is 7, since $4 \times 7 \equiv 1 \pmod 9$.

Multiplying both sides of the congruence by 7, we have:

$7 \times 4x \equiv 7 \times 5 \pmod 9$

$28x \equiv 35 \pmod 9$

Simplifying further:

$x \equiv 35 \equiv 8 \pmod 9$

Therefore, the solution to the congruence equation $4x \equiv 5$ (mod 9) is $x \equiv 8 \pmod 9$.

To solve the congruence equation $7x \equiv 8 \pmod 9$, we can find the multiplicative inverse of 7 modulo 9 and multiply both sides of the congruence by this inverse.

In this case, the multiplicative inverse of 7 modulo 9 is 4, since $7 \times 4 \equiv 1 \pmod 9$.

Multiplying both sides of the congruence by 4, we have:

$4 \times 7x \equiv 4 \times 8 \pmod 9$

$28x \equiv 32 \pmod 9$

Simplifying further:

$x \equiv 32 \equiv 5 \pmod 9$

Therefore, the solution to the congruence equation $7x \equiv 8$ (mod 9) is $x \equiv 5 \pmod 9$.

$3x \equiv 6 \pmod 9$ if and only if $3x = 9k + 6$, where $k \in \mathbb{Z}$. Solving for x, we have $x = 3k + 2$, where $k \in \mathbb{Z}$. $\square$

Exercise 11 (Medium):

prove that $n^3 + (n+1)^3 + (n+2)^3$ is divisible by 9 for any positive integer n.

Solution. we can create a table showing the values of the expression modulo 9 for different values of n. If all the values in the table are congruent to 0 modulo 9, then the expression is divisible by 9.

Let's create the table:

n	$n^3 + (n+1)^3 + (n+2)^3 \pmod 9$
1	0
2	0
3	0
4	0
5	0
6	0
7	0
8	0
9	0

In the table, we can see that for all values of n from 1 to 9, the expression $n^3 + (n+1)^3 + (n+2)^3$ evaluates to 0 modulo 9. Therefore, the expression is divisible by 9 for any positive integer n $\qquad\square$

Exercise 12 (easy):
Find out the remainder of the euclidean division of 5^n by 3.

Solution. $5^n \equiv (-1)^n \pmod 3$

If n is odd, the remainder is 2 ($5^n \equiv 2 \pmod 3$).

If n is even, the remainder is 1 ($5^n \equiv 1 \pmod 3$).

$\square$

Exercise 13 (Medium):
1/Find out the remainder of the euclidean division of 2^2012 by 5. 2/Find out the units digit of 2^2012.

Solution. 1/

$2^4 \equiv 1 \pmod 5$ since $2012 = 4*503+0$ then $2^{2012} = (2^4)^{503} \equiv 1^{503} \equiv 1 \pmod 5$ Therefore, the remainder of the Euclidean division of 2^2012 by 5 is 1.

2/

$2^{2012} \equiv 1 \pmod 5$ then the units digit of 2^2012 is 1 or 6.

However 2^2012 is even then the units digit is 6

$\square$

Exercise 14 (Hard):
A rep-unit is an integer which decimal representation only includes the digit 1 .

In this exercise we'll try to find out all rep-units which are perfect squares.

1/Given $n \in \mathbb{N}$, let's assume that the units digit of n^2 is 1 a) If $n^2 \equiv 1 \pmod{10}$, then $n \equiv -1 \pmod{10}$ or $n \equiv 1 \pmod{10}$. Therefore, the units digit of n can be either 1 or 10.

b) If $n^2 \equiv 1 \pmod{10}$, then $n = 10k + 1$ or $n = 10k - 1$, where $k \geq 1$. Hence, $n^2 = 100k^2 + 1 \pm 20k$. Therefore, $n^2 \equiv 1 \pmod{20}$.

2) Let's assume that there exists a rep-unit which is a perfect square. Then we have $11111\ldots11111 = n^2$, where $n \in \mathbb{N}$. Hence, $n^2 \equiv 11 \pmod{20}$, which is a contradiction. We have previously proved that $n^2 \equiv 1 \pmod{20}$. Thus, there are no rep-units which are perfect squares.

a/which values can the units digit of n take ?

b/prove that $n^2 \equiv 1 \pmod{20}$

2/find all rep-units which are perfect squares.

Solution. a) If $n^2 \equiv 1 \pmod{10}$, then $n \equiv -1 \pmod{10}$ or $n \equiv 1 \pmod{10}$. Therefore, the units digit of n can be either 1 or 10.

b) If $n^2 \equiv 1 \pmod{10}$, then $n = 10k + 1$ or $n = 10k - 1$, where $k \geq 1$. Hence, $n^2 = 100k^2 + 1 \pm 20k$. Therefore, $n^2 \equiv 1 \pmod{20}$.

2) Let's assume that there exists a rep-unit which is a perfect square. Then we have $11111\ldots11111 = n^2$, where $n \in \mathbb{N}$. Hence, $n^2 \equiv 11 \pmod{20}$, which is a contradiction. We have previously proved that $n^2 \equiv 1 \pmod{20}$. Thus, there are no rep-units which are perfect squares. $\square$

Exercise 15 (Hard):

1/Solve the equation (E):$11x^2 - 7y^2 = 5$ where $x, y \in \mathbb{Z}$

Solution. Assuming that there exists a solution (x, y) for the equation (E): $11x^2 - 7y^2 = 5$, we can rewrite it as:

$11x^2 \equiv 7y^2 + 5 \pmod{5}$.

Simplifying further, we have:

$11x^2 \equiv 7y^2 \pmod{5}$.

Let's construct the congruence tables for $x^2 \pmod 5$ and $2y^2 \pmod 5$:

$x^2 \pmod 5$	$2y^2 \pmod 5$
0	0
1	2
4	3
4	0
1	2

From the tables, we observe that $x^2 \equiv 2y^2 \pmod 5$ if and only if $x \equiv 0 \pmod 5$ and $y \equiv 0 \pmod 5$. Therefore, we can express x and y as:

$x = 5p$, where $p \in \mathbb{Z}$.

$y = 5q$, where $q \in \mathbb{Z}$.

Substituting these values back into the original equation, we get:

$11(25p^2) - 7(25q^2) = 5$.

Simplifying, we have:

$55p^2 - 35q^2 = 1$.

However, this equation is impossible because 55 and 35 are not coprime. Therefore, equation (E) has no solutions in $\mathbb{Z} \times \mathbb{Z}$.

$\square$

Exercise 16(Hard): Given $F_n = 2^{2^n} + 1$ $n \in \mathbb{N}$ what is the units digit of F_n for $n \geq 2$.

Solution. To determine the units digit of F_n for $n \geq 2$, let's observe the pattern and properties of powers of 2.

Let's calculate the units digits of F_n for various values of n:

$F_2 = 2^{2^2} + 1 = 2^4 + 1 = 16 + 1 = 17$ The units digit of F_2 is 7.

$F_3 = 2^{2^3} + 1 = 2^8 + 1 = 256 + 1 = 257$ The units digit of F_3 is 7.

$F_4 = 2^{2^4} + 1 = 2^{16} + 1 = 65536 + 1 = 65537$ The units digit of F_4 is 7.

We can observe that for $n \geq 2$, the units digit of F_n is always 7.

To prove this pattern by induction:

We start by checking the base case, which is $n = 2$. $F_2 = 2^{2^2} + 1 = 17$. The units digit of F_2 is 7, which satisfies the property.

Assume that the units digit of F_k is 7 for some positive integer $k \geq 2$. That is, assume F_k has a units digit of 7.

We need to prove that if the statement holds for k, it also holds for $k + 1$. That is, we need to show that the units digit of F_{k+1} is 7.

Let's consider $F_{k+1} = 2^{2^{k+1}} + 1$. Using the property that the units digit repeats in cycles of 4 for powers of 2, we can simplify the expression:

$2^{2^{k+1}}$ has the same units digit as 2^{2^k}.

From our inductive hypothesis, we know that F_k has a units digit of 7. Therefore, 2^{2^k} ends with a 7.

Adding 1 to a number ending in 7 results in a units digit of 8.

Therefore, the units digit of $F_{k+1} = 2^{2^{k+1}} + 1$ is 8.

By the principle of mathematical induction, we can conclude that the units digit of F_n is 7 for $n \geq 2$.

Hence, we have proven by induction that for $n \geq 2$, the units digit of F_n is 7.

$\square$

Exercise 17 (Hard):

Solve in $\mathbb{N} \times \mathbb{N}$ the equation $(E) : 2^m - 3^n = 1$.

Solution. Let's try to solve the problem modulo 8. Since $3^n \equiv 1$ mod 8 if n is even and $3^n \equiv 3 \mod 8$ if n is odd, then if $m \geq 3$, $2^m = 2^{m-3} \times 2^3 \equiv 0 \mod 8$. Hence, $3^n + 1 \equiv 0 \mod 8$, which means $2 \equiv 0 \mod 8$ and $4 \equiv 0 \mod 8$, which is impossible. Hence, $m \leq 2$. Therefore, we have 3 cases: $m = 0$, then $3^n = 0$ which is impossible; $m = 1$, then $3^n = 1$ which implies $n = 0$; $m = 2$, then $3^n = 3$ which implies $m = 1$. To sum up, there are only 2 solutions: $(2, 1)$ and $(1, 0)$.

$\square$

Exercise 18 (Very Hard):

Consider the sequence (U_n) defined by the following recurrence relation:
$$U_0 = 14$$
$$U_{n+1} = 5U_n - 6$$

Find the last two digits of U_n.

Solution. First of all, we will prove that $U_{n+2} \equiv U_n \mod 4$ for all $n \in \mathbb{N}$. $U_{n+2} = 5U_{n+1} - 6 = 5(5U_n - 6) - 6 = 25U_n - 36 \equiv U_n \mod 4$. Hence, for all $k \in \mathbb{N}$, $U_{2k} \equiv U_0 \mod 4 \equiv 2 \mod 4$ and $U_{2k+1} \equiv U_1 \mod 4 \equiv 64 \mod 4 \equiv 0 \mod 4$.

Next, we will prove by induction that $2U_n = 5^{n+2} + 3$. Let's call this statement P_n: "$2U_n = 5^{n+2}+3$ for all $n \geq 0$". For $n = 0$, $2U_0 = 28 = 5^{0+2}+3$. Let's assume P_n is true and prove that P_{n+1} is true. $2U_{n+1} = 10U_n - 12 = 5 \cdot 2U_n - 12 = 5 \cdot (5^{n+2} + 3) - 12 = 5^{n+3} + 15 - 12 = 5^{n+3} + 3$. Thus, P_{n+1} is true. By the induction principle, P_n is true. Therefore, $2U_n = 5^{n+2} + 3 \equiv 5^{n+2} + 3$ mod 100.

Now, consider $p \geq 2$, we have $5^p \equiv 25 \mod 100$. Since $n \geq 0$, $n + 2 \geq 2$, thus $2U_n \equiv 25 + 3 \mod 100 \equiv 28 \mod 100$.

As a result, for all $k \geq 0$, we have $2U_{2k} = 100s + 28$, which implies $U_{2k} = 50s + 14$. If $s = 2p$, then $U_{2k} = 100p + 14$, and we can conclude that $U_{2k} \equiv 14 \mod 100$. On the other hand, if $s = 2p + 1$, then $U_{2k} = 100p + 64 = 4(8 + 25p) \equiv 0 \mod 4$, which is impossible since $U_{2k} \equiv 2 \mod 4$. Therefore, we have $U_{2k} \equiv 14 \mod 100$.

Furthermore, consider $2U_{2k+1} = 28 + 100s$, which leads to $U_{2k+1} = 14 + 50s$. If $s = 2p$, then $U_{2k+1} = 14 + 100s \equiv 2 \mod 4$, which is impossible. If $s = 2p+1$, then $U_{2k} = 64 + 100p$, and we can conclude that $U_{2k+1} \equiv 64 \mod 100$.

Hence, the last two digits of (U_n) are 64 if n is odd and 14 if n is even. $\square$

Exercise 19 (Medium):
Solve in $\mathbb{Z} \times \mathbb{Z}$ the equation (E): $x^2 + 4y^2 = 3$.

Solution. To solve the equation $x^2 + 4y^2 = 3$ in $\mathbb{Z} \times \mathbb{Z}$, we can rewrite it as $x^2 \equiv -4y^2+3 \pmod 4$. Therefore, $x^2 \equiv 3 \pmod 4$.

Let's construct a congruence table modulo 4 for x^2:

x	$x^2 \pmod 4$
0	0
1	1
2	0
3	1

From the congruence table, we observe that x^2 is not congruent to 3 (mod 4). Therefore, the equation (E) has no solution in $\mathbb{Z} \times \mathbb{Z}$.

$\square$

Exercise 20 (Medium):

Solve in $\mathbb{Z}$ the following equation (E): $x^2 \equiv 25 \pmod{p}$, where p is a prime number.

Solution. Given the congruence $x^2 \equiv 25 \pmod{p}$, we can rewrite it as $x^2 - 25 \equiv 0 \pmod{p}$. This implies that $(x-5)(x+5) \equiv 0 \pmod{p}$. Since p is a prime number, we can conclude that either $(x-5) \equiv 0 \pmod{p}$ or $(x+5) \equiv 0 \pmod{p}$. Therefore, $x \equiv 5 \pmod{p}$ or $x \equiv -5 \pmod{p}$. In other words, $x = pk + 5$ or $x = pk' - 5$ for some integers k and k'.

$\square$

4 Diophantine equation

Exercise 1 (easy):

Solve in $\mathbb{Z} \times \mathbb{Z}$ the equation:

$$5x + 3y = 4$$

Solution. Since $\gcd(5,3) = 1$, this equation has solutions in $\mathbb{Z} \times \mathbb{Z}$. We know that $5 \cdot 2 + 3 \cdot (-2) = 4$. Thus, $5(x - 2) + 3(y - 2) = 0$. This implies $5(x - 2) = -3(y + 2)$. As $\gcd(5,3) = 1$, using Gauss's Lemma, we can conclude that 5 divides $-3(y+2)$. Hence, $-y - 2 = 5k$, where $k \in \mathbb{Z}$. Therefore, $y = -2 - 5k$, where $k \in \mathbb{Z}$. Substituting this into the equation, we get $5(x - 2) = -3(-5k) = 15k$. This implies $x = 3k + 2$. To verify, for $x = 3k + 2$ and $y = -5k - 2$, where $k \in \mathbb{Z}$, we have $5(3k + 2) + 3(-5k - 2) = 15k + 10 - 15k - 6 = 4$.

$\square$

Exercise 2 (Medium):

Solve in $\mathbb{Z} \times \mathbb{Z}$ the equation:

$$xy = 2x + 3y$$

Solution. We have the equation $xy = 2x + 3y$. Rearranging, we get

$(x - 3)(y - 2) = 6$. Hence, $x - 3$ divides 6. The possible values for $x - 3$ are $-6, -3, -2, -1, 0, 1, 2, 3, 6$. Thus, x can take the values $-3, 0, 1, 2, 3, 4, 5, 6, 9$. After checking all the cases, the solutions to the equation are

$(-3, 1), (0, 0), (1, -1), (2, -4), (4, 8), (5, 5), (6, 4), (9, 3)$.

$\square$

Exercise 3 (Medium):

Solve in $\mathbb{Z} \times \mathbb{Z}$ the equation:

$$x^2 - 5y^2 = 3$$

Solution. We have the equation $x^2 - 5y^2 = 3$. Simplifying, we get $x^2 \equiv 3 + 5y^2 \pmod{5}$. Let's create a congruence table modulo 5:

x	$x^2 \pmod 5$
0	0
1	1
2	4
3	4
4	1

From the table, we observe that x^2 is not congruent to 3 modulo 5 for any value of x. Therefore, the equation has no solutions. $\qquad\square$

Exercise 4 (easy): Solve the system of congruences in $\mathbb{Z}$:

$$n \equiv 8 \pmod{11}$$
$$n \equiv 2 \pmod{5}$$

Solution. To solve the system of congruences in $\mathbb{Z}$:

$$n \equiv 8 \pmod{11}$$
$$n \equiv 2 \pmod{5}$$

We can use the Chinese Remainder Theorem (CRT) to find the solution. First, let's examine each congruence individually:

For the congruence $n \equiv 8 \pmod{11}$, we can write it as $n = 11k + 8$ for some integer k.

Substituting this into the second congruence, we have $11k + 8 \equiv 2 \pmod 5$. Simplifying, we get $k \equiv 4 \pmod 5$.

Now, we can express k as $k = 5m + 4$ for some integer m.

Substituting this back into the first congruence, we have $n = 11(5m + 4) + 8$. Expanding, we get $n = 55m + 52$.

Thus, the general solution to the system of congruences is $n = 55m + 52$, where m is an integer.

verification : $n = 55m + 52$ satisfies the system.

$\square$

Exercise 5 (Medium): Solve the system of congruences in $\mathbb{Z}$:

$$\begin{cases} n \equiv 2 \pmod 3 \\ n \equiv 4 \pmod 7 \end{cases}$$

Solution. To solve the system of congruences in $\mathbb{Z}$, where $n \equiv 2 \pmod 3$ and $n \equiv 4 \pmod 7$, we can use the Chinese Remainder Theorem (CRT). First, let's examine each congruence individually:

For the congruence $n \equiv 2 \pmod 3$, we can write it as $n = 3k + 2$ for some integer k.

Substituting this into the second congruence, we have $3k + 2 \equiv 4 \pmod 7$. Simplifying, we get $3k \equiv 2 \pmod 7$.

To solve this congruence, we can find the modular inverse of 3 modulo 7. The modular inverse is 5, as $3 \cdot 5 \equiv 1 \pmod 7$.

Multiplying both sides of the congruence by 5, we get $k \equiv 10 \pmod 7$, which simplifies to $k \equiv 3 \pmod 7$.

Now, we can express k as $k = 7m + 3$ for some integer m.

Substituting this back into the first congruence, we have $n = 3(7m + 3) + 2$. Expanding, we get $n = 21m + 11$.

Thus, the general solution to the system of congruences is $n = 21m + 11$, where m is an integer.

verification : $n = 21m + 11$ satisfies the system.

$\square$

Exercise 6 (hard):

Solve the system of congruences in $\mathbb{Z}$

$$\begin{cases} n \equiv 2 \pmod{3} \\ n \equiv 6 \pmod{7} \\ n \equiv 1 \pmod{5} \end{cases}$$

Solution. To solve the system of congruences in $\mathbb{Z}$, where:

$$\begin{cases} n \equiv 2 \pmod{3} \\ n \equiv 6 \pmod{7} \\ n \equiv 1 \pmod{5} \end{cases}$$

We can use the Chinese Remainder Theorem (CRT). Let's examine each congruence individually:

For the congruence $n \equiv 2 \pmod{3}$, we can write it as $n = 3k + 2$ for some integer k.

Substituting this into the second congruence, we have $3k + 2 \equiv 6 \pmod{7}$. Simplifying, we get $3k \equiv 4 \pmod{7}$.

To solve this congruence, we can find the modular inverse of 3 modulo 7. The modular inverse is 5, as $3 \cdot 5 \equiv 1 \pmod{7}$.

Multiplying both sides of the congruence by 5, we get $k \equiv 20 \pmod{7}$, which simplifies to $k \equiv 6 \pmod{7}$.

Now, we can express k as $k = 7m + 6$ for some integer m.

Substituting this back into the first congruence, we have $n = 3(7m + 6) + 2$. Expanding, we get $n = 21m + 20$.

Now, let's consider the third congruence. Since $n \equiv 1 \pmod 5$, we have $21m + 20 \equiv 1 \pmod 5$. Simplifying, we get $m \equiv 1 \pmod 5$.

Finally, we can express m as $m = 5p + 1$ for some integer p.

Substituting this back into the expression for n, we have $n = 21(5p + 1) + 20$. Expanding, we get $n = 105p + 41$.

Thus, the general solution to the system of congruences is $n = 105p + 41$, where p is an integer. Verification : $n = 105p + 41$ satisfies the system.

$\square$

Exercise 7 (hard): In this Exercise, we'll associate each of the 26 letters of the french alphabet to the numbers 0,1,...,25. we encode a number x as follows : The coded number $c(x)$ is the remainder of the euclidean division of $41x + 37$ by 26. Decode the Word ITOT.

Solution. To decode each of the following letters:

I $\longleftrightarrow$ 8, T $\longleftrightarrow$ 19, O $\longleftrightarrow$ 14

We can use modular arithmetic and the concept of modular inverses. Let's begin with decoding the letter I. We have the congruence $41x + 37 \equiv 8 \pmod{26}$. Simplifying, we get $41x \equiv 23 \pmod{26}$.

To find the multiplicative inverse of 41 modulo 26, we need to find a number n such that $41n \equiv 1 \pmod{26}$. By solving the equation $41n - 26k = 1$, where k is an integer, we find that $n = 7$ satisfies the equation (since $41 \cdot 7 - 26 \cdot 11 = 1$).

Therefore, the multiplicative inverse of 41 modulo 26 is 7. Now, multiplying both sides of the congruence $41x \equiv 23 \pmod{26}$ by 7, we get $x \equiv 5 \pmod{26}$.

So, the letter I decodes to the letter F. Applying the same method, we can decode the letters T and O as well.

For T, we have $41x \equiv 19 \pmod{26}$. Multiplying both sides by 7, we get $x \equiv 15 \pmod{26}$. Therefore, T decodes to the letter E.

For O, we have $41x \equiv 14 \pmod{26}$. Multiplying both sides by 7, we get $x \equiv 2 \pmod{26}$. Therefore, O decodes to the letter V.

As a result, the decoded message is FEVE $\longleftrightarrow$ ITOT.

$\square$

Exercise 8 (Medium):
For what integers a and b does the equality $a^2 - 4ab = -4b^2 + 9$ hold?

Solution. We can see that $(x-y)^2 = x^2 - 2xy + y^2$. Also, we know that $a^2 - 4ab = -4b^2 + 9$ is equivalent to $a^2 - 2a(2b) + (2b)^2 = 9$, and can therefore conclude that $(a - 2b)^2 = 9$. The number 9 can be factored as $9 \cdot 1$, $3 \cdot 3$, $(-9) \cdot (-1)$, and $(-3) \cdot (-3)$. Because the left side of the equation is the product of two identical integers $(a - 2b)$, we can conclude that $a - 2b = \pm 3$, giving us $a = \pm 3 + 2b$.

$\square$

Exercise 9 (hard):
The positive integers a, b, c are pairwise relatively prime, a and c are odd, and the numbers satisfy the equation $a^2 + b^2 = c^2$. Prove that $b + c$ is a square of an integer.

Solution. We can rewrite the equation in a more convenient form as $a^2 = c^2 - b^2$. Factoring the right-hand side using identity (2) yields $a^2 = (c+b)(c-b)$. As previously explained, it is sufficient to prove that $b + c$ and $c - b$ are relatively prime for $b + c$ and $c - b$ to be squares. Let us assume the opposite, that $b + c$ and $c - b$ share a divisor $d > 1$. Let $b + c = dx$ and $c - b = dy$, where x and y are integers. Adding the equations together gives $2c = d(x + y)$, and subtracting them yields $2b = d(x - y)$. This means that either d divides both b and c, or 2 divides d. We can exclude the first case since b and c are relatively prime. If d were even, b and c would have had the same parity, which would in turn imply that a is even, resulting in a contradiction. This means that we can exclude the second case as well. Hence, $c + b$ and $c - b$ must be relatively prime, meaning that $b + c$ is the square of an integer (which is also true for $c - b$), which was to be proven.

$\square$

Exercise 10 (hard): Find all integer solutions to the equation $x^3 + y^3 - 3xy = 3$.

Solution. Trying to immediately factor the expression in the left hand side will not lead to much progress. Nevertheless, the expression closely resembles the left hand side of identity (4), where z is replaced by 1. Comparing the expressions $x^3 + y^3 - 3xy$ and $x^3 + y^3 + z^3 - 3xyz$ where $z = 1$, we see that they only differ by the number 1. Therefore, we can add 1 to both sides of the equation $x^3 + y^3 - 3xy = 3$ to make the left hand side factorizable. Consequently, using identity (4), the obtained equation $x^3 + y^3 + 1 - 3xy = 4$ can be rewritten as

$$(x + y + 1)(x^2 + y^2 + 1 - xy - x - y) = 4.$$

There are six possible ways to split up the number 4 between the two factors, namely

$$(x+y+1, x^2+y^2+1-xy-x-y) = (1,4), (4,1), (2,2), (-1,-4), (-4,-1)$$

or $(-2, -2)$. For the first factor of the left hand side in equation (5) to be even, x and y must have different parity, making the second factor odd. Hence, we can exclude the alternatives $(2, 2)$ and $(-2, -2)$. The other alternatives, i.e.

$$(x + y + 1, x^2 + y^2 + 1 - xy - x - y) = (1,4), (4,1), (-1,-4)$$

or $(-4, -1)$, lead to the systems of equations

$$\begin{cases} x + y = 0, \\ \quad xy = -1, \end{cases}$$

$$\begin{cases} x + y = 3, \\ \quad xy = 2, \end{cases}$$

$$\begin{cases} x + y = -2, \\ \quad 3xy = 11, \end{cases}$$

$$\begin{cases} x + y = -5, \\ \quad 3xy = 32, \end{cases}$$

respectively, which follows from rewriting the second factor as $(x+y)^2+1-3xy-(x+y)$. The two latter can be excluded, since

11 and 32 are not divisible by 3. The two first equations give, by substituting the upper equation into the lower, the solutions $(x, y) = (1, -1), (-1, 1), (1, 2)$ and $(2, 1)$. These are hence the integer solutions to the equation.

$\square$

TO BE CONTINUED ...